STUDENT GUIDE

DESIGNING SPACES

VISUALIZING, PLANNING, AND BUILDING

MathScape
SEEING AND THINKING
MATHEMATICALLY

DESIGNING SPACES, INC.

Welcome Designing Spaces, Inc. Consultants!

Designing a house involves the mathematics of geometry and spatial visualization. To make sure that the homes appeal to people from different cultures, it is also important to understand how people from around the world use construction shapes to build structures that reflect their own cultural experiences.

How can you describe houses from around the world?

DESIGNING
SPACES

PHASE**ONE**
Visualizing and Representing Cube Structures

Your job as a consultant for Designing Spaces, Inc. will begin with exploring different ways to represent three-dimensional structures. In this first phase, you will build houses made from cubes. Then you will create building plans of a house. The true test of your plans will be whether another person can follow them to build the house. The skills you develop in this phase are an important basis for understanding geometry.

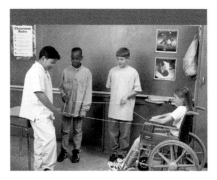

PHASE**TWO**
Functions and Properties of Shapes

What shapes do you see in the houses pictured on these pages? In this phase, you will explore the properties of two-dimensional shapes. You will apply what you learn to decode the formulas for making the construction shapes developed by the Designing Spaces, Inc. technical research group. At the end of the phase, Designing Spaces, Inc. will turn to you for advice about the dimensions of their construction shapes.

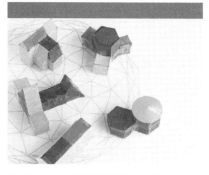

PHASE**THREE**
Visualizing and Representing Polyhedrons

In Phase Three, you will explore how you can use the building shapes you helped design in Phase Two to make a variety of three-dimensional structures. You'll learn names for the structures you create and make drawings of them. The phase ends with a final project. You will use the building shapes to design a home for a cold and snowy climate, or a warm and rainy climate. Then you'll create building plans for your design.

PHASE ONE

To: House Designers
From: General Manager
Designing Spaces, Inc.

Welcome to Designing Spaces, Inc.! In your new job as a house designer, you will be asked to design different kinds of homes for our customers.

One kind of home our company designs is a low-cost modular home. These homes are made from cube-shaped rooms that are all the same size. We can make many different kinds of modular homes because the rooms fit together in so many different ways.

In Phase One, you will take on the special assignment of designing houses made of cube-shaped rooms. You will learn helpful ways to make building plans for your structures. Your plans must be clear. Someone else should be able to build your house from your plans.

Visualizing and Representing Cube Structures

WHAT'S THE MATH?

Investigations in this section focus on:

MULTIPLE REPRESENTATIONS

- Represent three-dimensional structures with isometric and orthogonal drawings.

PROPERTIES and COMPONENTS of SHAPES

- Identify two-dimensional shapes that make up three-dimensional structures.

- Describe properties of structures in writing so that someone else can build the structure.

VISUALIZATION

- Build three-dimensional structures from two-dimensional representations.

MathScape Online
mathscape1.com/self_check_quiz

 # Planning and Building a Modular House

CREATING
TWO-DIMENSIONAL
REPRESENTATIONS

A modular house is made of parts that can be put together in different ways. You can use cubes to design a modular house model. Then you can record your design in a set of building plans. These plans are a way of communicating about your house design so that someone else could build it.

Use Cubes to Design a House

What kind of structure can you make from cubes that meets the design guidelines?

One kind of home that Designing Spaces, Inc. designs is a low-cost modular home made from cube-shaped rooms. For your first assignment, use eight to ten cubes to design a modular house model. Let each cube represent a room in the house. Follow the Design Guidelines for Modular Houses.

Cubes in the structure you design can be arranged like these.

 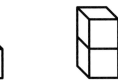

Cubes cannot be arranged like these.

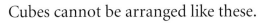

Design Guidelines for Modular Houses

- A face of one cube must line up exactly with a face of at least one other cube so that the rooms can be connected by stairways and doorways.

- No rooms can defy gravity. Each cube must rest on the desktop or directly on top of another cube.

Create Building Plans

Now that you have designed and built your modular house model, make a set of plans that someone else could use to build the same structure. Your plans should include the following:

- Create at least one drawing of the house.

- Write a description of the steps someone would follow to build the house.

How can you make two-dimensional drawings of three-dimensional structures?

Add to the Visual Glossary

In this unit, you will create your own Visual Glossary of geometric terms that describe shapes. These terms will help you to share your building plans with others. One example is the term *face*. The term is used in the Design Guidelines for Modular Houses.

- After your class discusses the meaning of the term *face*, add the class definition to your Visual Glossary.

- Include drawings to illustrate your definition.

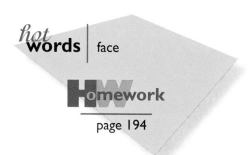

hot **words** | face

Homework
page 194

2 Seeing Around the Corners

REPRESENTING
THREE DIMENSIONS
IN ISOMETRIC
DRAWINGS

How many different houses do you think you can make with three cubes? How many can you make with four cubes? As you explore the possibilities, you will find that a special type of drawing called *isometric drawing* can help you record the different structures you make.

How many different structures is it possible to build with three cubes?

Build and Represent Three-Room Houses

Design as many different modular houses as you can using three cubes.

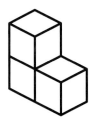 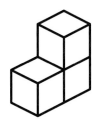

These two houses are the same. You can rotate one to be just like the other, without lifting it.

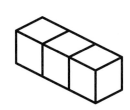 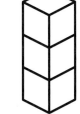

These two houses are different. You have to lift one house to make it just like the other.

- Follow the Design Guidelines for Modular Houses presented on page 166.

- Make an isometric drawing to record each different structure you make.

Sample Isometric Drawings

Isometric drawings show three faces of a structure in one sketch.

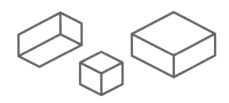

You can use isometric dot paper to help you in making isometric drawings.

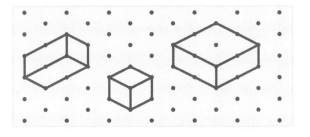

Build and Represent Four-Room Houses

You have explored the number of houses it is possible to build with three cubes. Now try using four cubes to design as many different houses as you can.

- Follow the Design Guidelines for Modular Houses presented on page 166.

- Record each different structure you make as an isometric drawing or another type of drawing.

How many different houses can you make with four cubes?

How many different structures is it possible to build with four cubes?

Add to the Visual Glossary

Think about the isometric drawings you made in this lesson. Look at the illustrations shown on this page. Then compare the lengths of the sides and the sizes of the angles in both kinds of drawings.

- After discussing with the class, add the class definition of the term *isometric drawing* to your Visual Glossary.

- Be sure to include drawings to illustrate the definition.

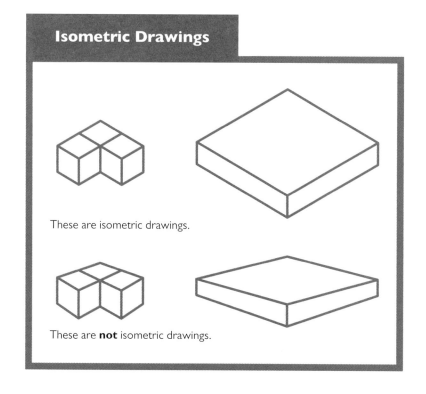

Isometric Drawings

These are isometric drawings.

These are **not** isometric drawings.

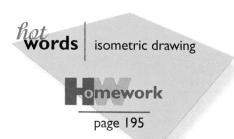

hot **words** | isometric drawing

Homework
page 195

3 Seeing All Possibilities

TRANSLATING
BETWEEN
ORTHOGONAL AND
ISOMETRIC
DRAWINGS

You have explored using isometric drawings to show three-dimensional structures on paper. Here you will try another drawing method called *orthogonal drawing*. You need to understand both types of drawings, so you can read building plans and create plans of your own.

Construct Houses from Orthogonal Drawings

How can you use orthogonal drawings to build three-dimensional structures?

Four sets of building plans for modular houses are shown here. Each plan is represented with orthogonal drawings that show three views of the house. Your job is to build each house with cubes and record the least number of cubes you needed to build the house.

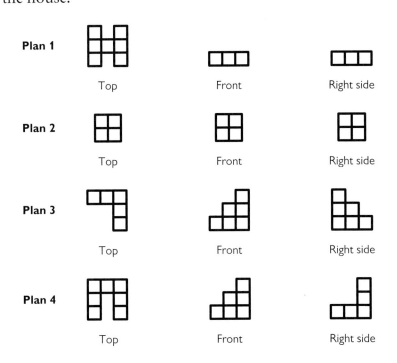

Plan 1 — Top, Front, Right side

Plan 2 — Top, Front, Right side

Plan 3 — Top, Front, Right side

Plan 4 — Top, Front, Right side

Make Orthogonal Drawings

These plans are isometric drawings of several houses. Your job is to make orthogonal drawings showing the top, front, and right side of each house.

How can you make orthogonal drawings from an isometric drawing of a house?

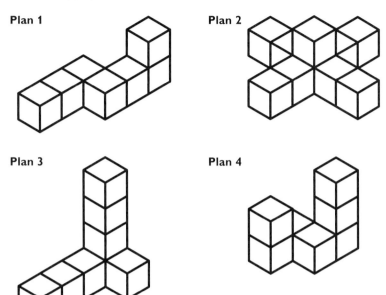

Plan 1

Plan 2

Plan 3

Plan 4

Add to the Visual Glossary

Think about the orthogonal drawings you made. Look at the illustrations showing drawings that are orthogonal and drawings that are not orthogonal.

- After class discussion, add the class definition of the term *orthogonal drawing* to your Visual Glossary.

- Be sure to include drawings to illustrate the definition.

Orthogonal Drawing

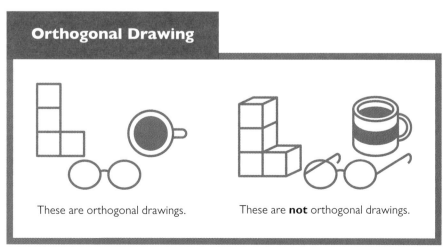

These are orthogonal drawings.

These are **not** orthogonal drawings.

hot **words** | orthogonal drawing

Homework
page 196

4 Picture This

You have learned how to make isometric and orthogonal drawings. In this lesson, you will use what you have learned to improve the building plans you made for your first assignment. One way to check how well your plans communicate is to get someone else's comments on your plans.

Evaluate and Improve Building Plans

How can you apply what you know to improve how well your plans communicate?

1 Evaluate your building plans from Lesson 1. Think about these questions and make changes to improve your building plans.

   a. Can you tell how many cubes were used in each structure? How do you know?

   b. Can you tell how the cubes should be arranged? How would you make the drawings clearer?

   c. Do the written plans include a step-by-step description of the building process? Are any steps missing?

   d. What are one or two things that you think are done well in the plans? What are one or two things that you think need to be changed in the plans?

2 Exchange your plans with a partner. Carefully follow your partner's plans and build the structure. Write down any suggestions that you think would improve your partner's building plans.

   a. Are the drawings clear? How could you make them clearer?

   b. Are the step-by-step written instructions easy to follow? What suggestions can you give to improve the instructions?

   c. What are one or two things that you think are done well in the plans?

Use Feedback to Revise Building Plans

Carefully review the feedback you received from your partner. Use it to make final changes to your plans.

- Keep in mind that feedback is suggestions and opinions that can help you improve your work.

- It is up to you to decide whether to use the feedback and how to use it.

How can you improve your plans so that they are easier for someone else to use?

Write About the Changes

When you finish changing your building plans, write a memo to the General Manager of Designing Spaces, Inc. The memo should summarize the revisions you made.

- Describe the changes you made in your drawings.

- Describe the steps it took to build your structure.

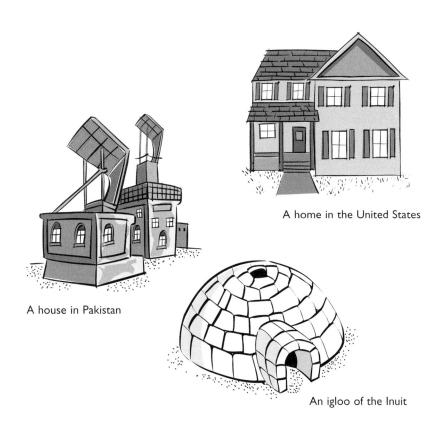

A home in the United States

A house in Pakistan

An igloo of the Inuit

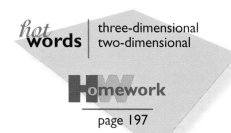

hot words | three-dimensional two-dimensional

Homework

page 197

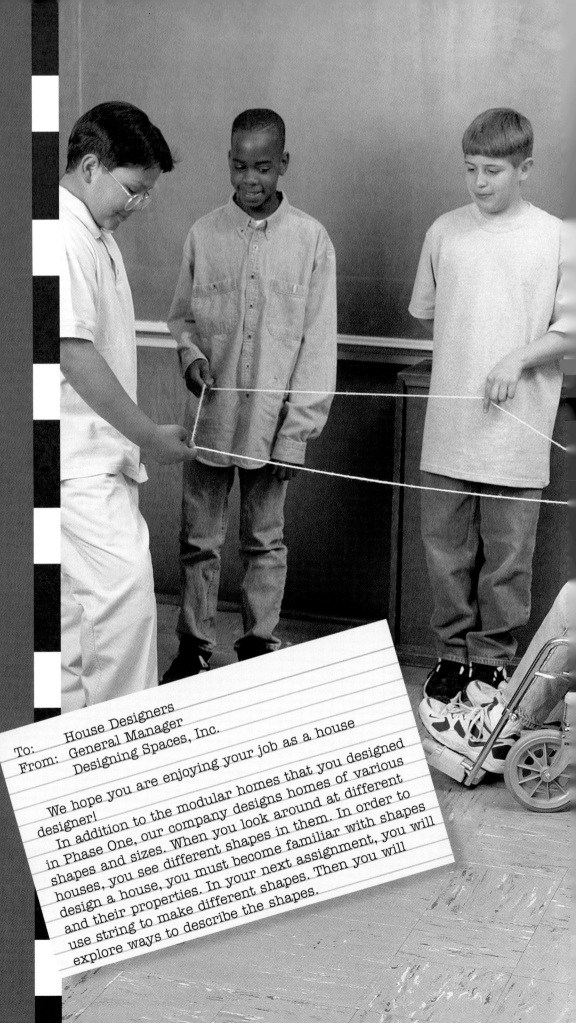

PHASE TWO

In Phase Two, you will investigate various shapes that can be used to design and build houses. You will learn about the properties of shapes and how to measure angles with a protractor—ideas that are good to know when constructing a house.

Functions and Properties of Shapes

WHAT'S THE MATH?

Investigations in this section focus on:

PROPERTIES and COMPONENTS of SHAPES

- Identify two-dimensional shapes and their properties.

- Measure sides and angles of shapes.

- Estimate area and perimeter of shapes.

- Use geometric notation to indicate relationships between angles and sides in shapes.

- Describe properties of shapes in writing.

- Expand vocabulary for describing shapes.

VISUALIZATION

- Visualize shapes from clues about their sides and angles.

- Perform visual and mental experiments with shapes.

MathScape Online
mathscape1.com/self_check_quiz

5 String Shapes

What shapes can you find in the houses in your neighborhood? Shapes that have three or more sides are called **polygons.** As you make shapes from string, you will learn some special properties of sides of polygons. You'll also learn mathematical names for the shapes you create.

Can you find a shape that satisfies clues about parallel and equilateral sides?

Make Shapes from Clues

Do the following for each shape clue given:

1 Try making a shape that fits the description in the clue.

2 If you can make the shape, record it. Then label the equal sides and the parallel sides. (The top of the handout Naming and Labeling Polygons will help you with this.) If you can't make the shape, write "Impossible."

3 Label the shape you draw with a mathematical name. If you don't know a name for the shape, make up a name that you think describes the properties of the shape. (The bottom of the handout Naming and Labeling Polygons will help you with this.)

Shape Clue 1: An equilateral shape with more than 3 sides and no parallel sides.	**Shape Clue 2:** A shape with 2 sides equal and 2 different sides parallel but not equal. (Hint: Your shape can have more than 4 sides.)
Shape Clue 3: A quadrilateral with 2 pairs of parallel sides and only 2 sides equal.	**Shape Clue 4:** A shape with at least 2 pairs of parallel sides that is not an equilateral shape. (Hint: Your shape can have more than 4 sides.)
Shape Clue 5: A quadrilateral that is equilateral and has no parallel sides.	**Shape Clue 6:** Make up a clue of your own and write it down. Test it to see if you can make the shape, and write a sentence telling why you can or cannot make it.

Make Animated Shapes

Animated shapes are shapes that a group makes with string. The group starts with one shape, and then one member of the group changes positions to change the shape as the rest stand still. Make each of the animated shapes on the handout Animated Shapes.

1 Try to make the shape by moving the fewest people.

2 Record how you made the shape with drawings or words. Include your starting positions, who moved, and where.

Using string, how can you change a square into a triangle? a trapezoid into a square?

Add to the Visual Glossary

Think about how you used the terms *parallel* and *equilateral* to describe the sides of shapes. Then look at the illustrations showing lines and walls that are parallel and lines and walls that are not parallel.

- After some class discussion, write the class's description of what is meant by the terms *parallel* and *equilateral* in your Visual Glossary.

- Be sure to include drawings to illustrate your description.

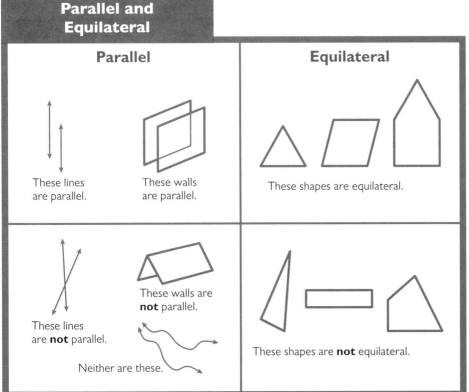

Parallel and Equilateral

Parallel	Equilateral
These lines are parallel. These walls are parallel.	These shapes are equilateral.
These lines are **not** parallel. These walls are **not** parallel. Neither are these.	These shapes are **not** equilateral.

hot **words** | parallel
equilateral

Homework
page 198

6 Polygon Paths

Where the sides of a shape meet, they form angles.
Building plans for houses have to show the measures of angles and the lengths of sides. Here you will explore how measuring angles is different from measuring lengths. Then you will use what you have learned to describe polygons.

Measure Angles

How can you tell if two angles are equal?

Use a protractor to measure the angles in this polygon. Be sure to refer to the guidelines on How to Use a Protractor.

- Which is the smallest angle? How did you measure it?

- Which is the largest angle? How did you measure it?

- Are any of the angles equal?

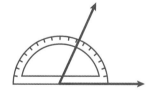

How to Use a Protractor

- Place the 0° line along one side of the angle.

- Center the protractor where the sides meet at the vertex.

- Read the degree mark that is closest to where the other side of the angle crosses the protractor.

TIP: If the side of an angle doesn't cross the protractor, imagine where it would cross if the line was longer. Or copy the figure and extend the line.

Write Polygon Path Instructions

Polygon path instructions describe the steps you take when drawing a polygon. Study the example shown. Then write polygon path instructions for the shapes on the handout.

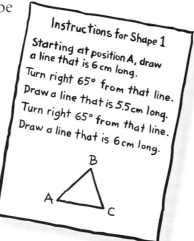

Instructions for Shape 1
Starting at position A, draw a line that is 6 cm long.
Turn right 65° from that line.
Draw a line that is 5.5 cm long.
Turn right 65° from that line.
Draw a line that is 6 cm long.

How can you use measures of sides and angles to describe a polygon?

Add to the Visual Glossary

Think about how you have used the terms *equal angles* and *right angles* to describe angles in this lesson. Study the pictures of equal angles and right angles.

- After the class discussion, write in your Visual Glossary the meanings of *equal angles* and *right angles* that your class developed.

- Include drawings to illustrate the definitions.

Types of Angles

Equal Angles	Right Angles
These are equal angles. These are equal angles.	These angles are right angles.
These are **not** equal angles. These are **not** equal angles.	These angles are **not** right angles.

hot **words** | equal angles
right angles

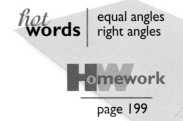

page 199

7 Shaping Up

In Lesson 5, you made string shapes from clues about sides. Here, you will make shapes from clues about angles. Then you will apply what you've learned so far in the Sides and Angles Game.

Make Shapes from Clues

How can you use what you know about the properties of sides and angles to make shapes?

Do the following for each shape clue given:

1️⃣ Try making a shape that fits the description in the clue.

2️⃣ If you can make the shape, record it. Then label the equal angles and the right angles. (The top of the handout Naming and Labeling Polygons will help you with this.) If you can't make the shape, write "Impossible."

3️⃣ Label the shape you draw with a mathematical name. If you don't know a name for the shape, make up a name that you think describes the properties of the shape. (The bottom of the handout Naming and Labeling Polygons will help you with this.)

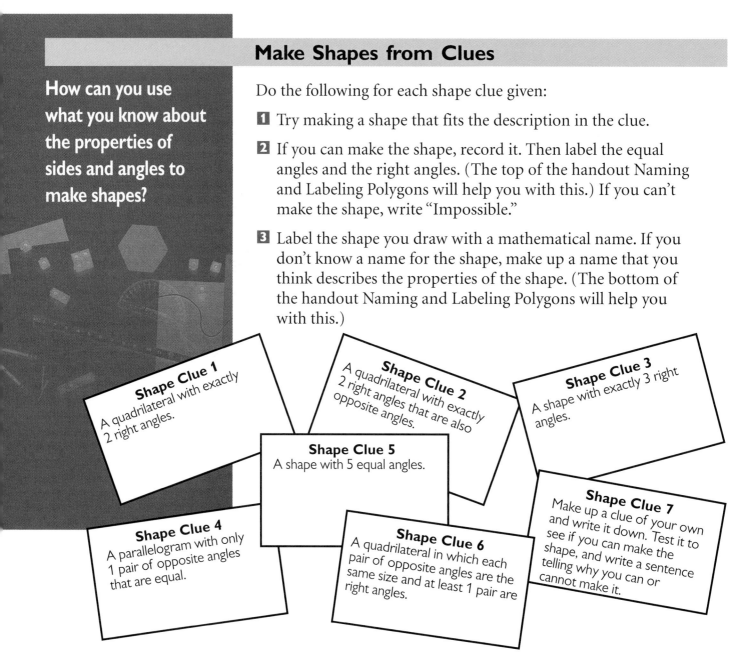

Shape Clue 1
A quadrilateral with exactly 2 right angles.

Shape Clue 2
A quadrilateral with exactly 2 right angles that are also opposite angles.

Shape Clue 3
A shape with exactly 3 right angles.

Shape Clue 5
A shape with 5 equal angles.

Shape Clue 4
A parallelogram with only 1 pair of opposite angles that are equal.

Shape Clue 6
A quadrilateral in which each pair of opposite angles are the same size and at least 1 pair are right angles.

Shape Clue 7
Make up a clue of your own and write it down. Test it to see if you can make the shape, and write a sentence telling why you can or cannot make it.

Play the Sides and Angles Game

In the Sides and Angles Game, you will try to make shapes that fit two different descriptions. One description is of the sides of the shape. The other is of the angles. Be sure to read the rules carefully before beginning.

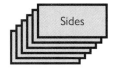 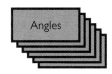

What shapes can you make from clues about sides and angles?

The Sides and Angles Game Rules

Before beginning the game, place the Sides cards face-down in one pile. Place the Angles cards face-down in another pile.

1. The first player takes one card from each pile.

2. The player tries to draw a shape that fits the descriptions on the two cards. Then the player marks the drawing to show any parallel sides, equal sides, equal angles, or right angles. The player should label the shape with a mathematical name or create a name that fits. If the player cannot make a shape, the player should describe why.

3. Players take turns trying to find another way to draw the shape, following the directions in Step 2. When the group draws four shapes or runs out of possible drawings, the next player picks two new cards. Play begins again.

Add to the Visual Glossary

Think how you have used the terms *regular shape* and *opposite angles* to describe the shapes you made in this lesson.

- After class discussion, write the class definition for each term in your Visual Glossary.

- Include drawings to illustrate the definitions.

Shapes and Angles

These are regular shapes.   These are opposite angles.

These are **not** regular shapes.   These are **not** opposite angles.

 hot **words** | regular shape / opposite angles

 Homework

page 200

8 Assembling the Pieces

You've explored sides, angles, and shape names when describing two-dimensional shapes. In this lesson, you will investigate two more properties of shapes—perimeter and area. Then you will apply everything you've learned to describe six shapes. You will use the shapes in the next phase to build a model home.

Investigate Perimeter and Area

How many shapes can you draw with a perimeter of 10 cm? 15 cm?

Perimeter and *area* are properties of shapes. **Perimeter** is the distance around a shape. **Area** is the number of square units a shape contains. Follow the directions given to investigate these two properties. You'll need a ruler for this activity.

The area of the small rectangle is 12 cm². The perimeter is 14 cm. Can you estimate the area and perimeter of the shaded triangle?

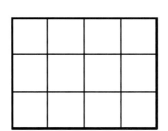 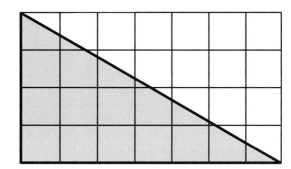

1 Choose a perimeter between 8 cm and 16 cm. Make at least four different shapes with that perimeter. Then record each shape you made and label the lengths of the sides.

2 Estimate the area in square centimeters of some of the shapes you recorded. You may find it helpful to trace the shapes on centimeter grid paper.

As you were drawing different shapes with the same perimeter, what did you notice about the area of the shapes? Why do you think this is true?

Describe Construction Shapes

Later in the unit, you will use six shapes to design and build a house. Your teacher will give you samples of these six shapes.

- Pretend that you are explaining to a manufacturer the shapes and the sizes you will be using.

- Describe each shape in as many ways as you can.

- Be sure to refer to the assessment criteria when writing your descriptions.

How can you use what you've learned to describe a shape so that someone else could draw it?

hot **words** | perimeter | area

HW**omework**

page 201

PHASE THREE

To: House Designers
From: General Manager
Designing Spaces, Inc.

Designing Spaces, Inc. has just been hired to create a new collection of house designs. Your work on the next few assignments leads up to designing one of these houses. Your house will be located in one of two climates: hot and rainy or cold and snowy. This project will give us a chance to see how much you've learned while working for Designing Spaces, Inc.

In Phase Two, you explored the properties of two-dimensional shapes that make up three-dimensional structures. In this phase, you will investigate the properties of the three-dimensional structures themselves. Using this information in the final project, you will design, build, and make plans for a model home!

Visualizing and Representing Polyhedrons

WHAT'S THE MATH?

Investigations in this section focus on:

MULTIPLE REPRESENTATIONS

- Use perspective drawing techniques to represent prisms and pyramids.
- Identify orthogonal views of a structure.

PROPERTIES and COMPONENTS of SHAPES

- Identify prisms and pyramids.
- Investigate the properties of edge, vertex, and face.
- Explore the relationships among the properties of two- and three-dimensional shapes.

VISUALIZATION

- Form visual images of structures in your mind from clues given about the structure.

MathScape Online
mathscape1.com/self_check_quiz

Beyond Boxes

MOVING FROM
POLYGONS TO
POLYHEDRONS

In Phase One, you designed houses with cubes. If you look at houses around the world, however, you will see many different shapes. Here you will use the shapes you made in Lesson 8 to build three-dimensional structures.

Construct Closed Three-Dimensional Structures

What three-dimensional structures can you build by choosing from six different shapes?

A **polyhedron** is any solid shape whose surface is made up of polygons. Using the shapes from Lesson 8, you will build three-dimensional models of polyhedrons.

1 Use the Shape Tracers to trace four of each of the following shapes on heavy paper: triangles, rectangles, squares, rhombi, trapezoids, and hexagons. Then carefully cut out the shapes.

2 Build at least two different structures from the shapes you cut out. Be sure to follow the Building Guidelines.

A thatched roof house in Central America

Building Guidelines

- Use 3–15 pieces for each structure that you build. If you need more pieces, cut them out.

- The base, or bottom, of the structure can be made of only **one** piece.

- The pieces must **not** overlap.

- The structure must be **closed.** No gaps are allowed. (Use tape to hold the shapes together.)

- No hidden pieces are allowed; you must be able to see them all.

Record and Describe Properties of Structures

In Phase Two, you used properties of sides and angles to describe polygons. You can use properties of faces, vertices, and edges to describe polyhedrons.

What are the properties of the three-dimensional structures you built?

1 For each structure you made, create a table of properties. Record the following numbers:

    a. faces   **b.** vertices   **c.** edges

    d. sets of parallel faces   **e.** sets of parallel edges

2 Write a description of one of the structures you built. Include enough information so that your classmates would be able to pick out your structure from all the others. You can include drawings. Think about the following when you write your description:

    a. What shapes did you use in the structure? What shape is the base?

    b. How many faces, vertices, and edges does the structure have?

    c. Are any edges parallel to each other?

    d. Are any faces parallel to each other?

Table of Properties

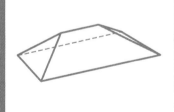

Shape of Base	Shapes Used	No. of Faces	No. of Vertices (Corners)	No. of Edges	Sets of Parallel Faces?	Sets of Parallel Edges?
Rectangle	1 rectangle 2 triangles 2 trapezoids	5	6	9	No (0)	Yes (2)

Add to the Visual Glossary

In this lesson, you used the terms *vertex* and *edge* to describe properties of polyhedrons. Write definitions for vertex and edge.

- After class discussion, add the class definitions of the terms *vertex* and *edge* to your Visual Glossary.

- Be sure to add drawings to illustrate your definitions.

hot **words** | vertex
edge

Homework

page 202

10 Drawing Tricks

VISUALIZING AND
DRAWING PRISMS
AND PYRAMIDS

Architects draw many different views of their house designs. In Phase One, you learned to create isometric drawings of houses made of cubes. The methods you learn here for drawing prisms and pyramids will prepare you for drawing house plans that include other shapes.

Draw Prisms

How can you draw a prism in two dimensions?

A **prism** has two parallel faces that can be any shape. These are its **bases.** A prism gets its name from the shape of its bases. For example, if the base is a square, it is called a square prism. Read the tips on How to Draw a Prism.

- Draw some prisms on your own.

- Label each prism you draw with its name.

How to Draw a Prism

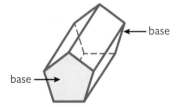

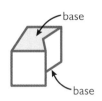

1. Draw the base of the prism. The example shown is the base for a pentagonal prism.

2. Draw the second base by making a copy of the first base. Keep corresponding sides parallel.

3. Connect the corresponding vertices of the two bases. This will produce a collection of parallel edges with the same length.

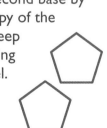

Draw Pyramids

In a **pyramid,** the base can be any shape. All the other faces are triangles. A pyramid gets its name from the shape of its base. For example, if the base is a triangle, it is called a triangular pyramid. Read the tips for How to Draw a Pyramid.

How can you draw a pyramid in two dimensions?

- Draw some pyramids on your own. Try some with different shapes as their bases.

- Label each pyramid you draw with its name.

How to Draw a Pyramid

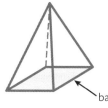 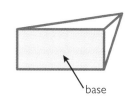
base   base

1. Draw the base. The example shown is the base for a pentagonal pyramid.

2. Mark any point outside the base.

3. Draw line segments from each vertex of the base to the point.

Add to the Visual Glossary

Think about the drawings you made of pyramids and prisms. What do all pyramids have in common? What do all prisms have in common?

- After class discussion, add the class definitions of the terms *prism* and *pyramid* to your Visual Glossary.

- Be sure to add drawings to illustrate your definitions.

hot **words** | prism
pyramid

Homework
page 203

11 Mystery Structures

Have you ever solved a puzzle from a set of clues? That's what you will do to build three-dimensional Mystery Structures. Then you will build your own Mystery Structure and write clues to go with it. To create an answer key for your clues, you will use the drawing skills you have learned.

Solve the Mystery Structures Game

How can you apply what you know about two-dimensional shapes to solve clues about three-dimensional structures?

The Mystery Structures Game pulls together all of the geometric concepts you have learned in this unit. Play the Mystery Structures Game with your group. You will need the following: Round 1 clues, a set of shapes (4 each of the triangle, rhombus, trapezoid, and hexagon; 6 each of the rectangle and square), and tape.

The Mystery Structures Game

How to play:

1. Each player reads, but does not show, one of the clues to the group. If your clue has a picture on it, describe the picture.

2. Discuss what the structure might look like.

3. Build a structure that matches all the clues. Recheck each clue to make sure the structure satisfies each one. You may need to revise the structure several times.

4. Make a drawing of the structure that shows depth.

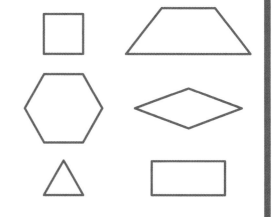

Create Clues

Now that you know how the Mystery Structures Game works, follow these steps to write your own set of clues for the game:

1 Build a closed, three-dimensional shape with up to 12 shape pieces. Make a structure that will be an interesting project for someone else to build.

2 Write a set of four clues about your structure. Write each clue on a separate sheet of paper. Use at least one orthogonal drawing. From your four clues, someone else should be able to build your structure.

3 Make a drawing of the structure showing depth to serve as an answer key for other students who try to follow your clues.

How can you use what you have learned about shapes to describe three-dimensional structures?

Give Feedback on Clues

Exchange clues with a partner. Then try to build your partner's structure. When you are done, compare your structure with the answer key. Write feedback on how your partner's clues might be improved.

- Are the drawings clear? How could you make them clearer?

- Are the clues easy to follow? What suggestions can you give to improve the clues?

- What are one or two things that you think are done well?

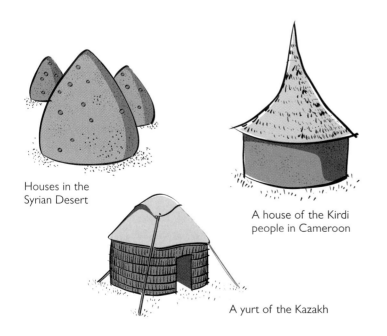

Houses in the Syrian Desert

A house of the Kirdi people in Cameroon

A yurt of the Kazakh

hot **words** | polygon
polyhedron

H**omework**

page 204

12 Putting the Pieces Together

The climate in which a house is built can influence the shape of the house and the materials used to build the house. For your final project, you will research houses in different climates around the world. Then you will apply what you have learned to design a house model for a specific climate.

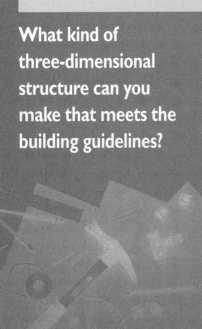

Build a House

What kind of three-dimensional structure can you make that meets the building guidelines?

Before you start building your house, read about home designs in different climates. Choose the climate for your house model: a hot and rainy climate or a cold and snowy climate. Think about the design features that your house should have for that climate. Then build the house model.

1 Use the Shape Tracers to trace and cut out a set of building shapes.

2 Build a house using the shape pieces. Be sure to follow the Building Guidelines when constructing your house.

Building Guidelines

- The home must have a roof. The roof should be constructed carefully so that it will not leak.

- The home must be able to stand on its own.

- You must use 20–24 shape pieces. You do not need to use each type of shape.

- You may add one new type of shape.

Create a Set of Plans

After building your house, make a set of plans. Another person should be able to use the plans to build your house. Use the following guidelines to help you create your plans:

How can you represent your house in two dimensions?

- Include both orthogonal and isometric drawings of your house. Be sure to label the drawings. The labels will make your plans easier to understand.

- Include a written, step-by-step description of the building process so that someone else could build your house. Use the names and properties of shapes to make your description clear and precise.

Write Design Specifications

Write a memo to the Directors of Designing Spaces, Inc. describing your house. The specifications should explain the design clearly. The Marketing Department should be able to understand and sell the design. The designers should be able to make the shapes and calculate the approximate costs. The memo should answer the following questions:

- What does your home look like?

- How would you describe the shape of the entire house?

- How would you describe the shape of its base, its roof, and any special features?

- What climate is your house designed for?

- What features does your home have to make it well suited to the climate?

- Where in the world could your house be located?

- What shapes are used in your house?

- How many of each shape are there?

- How many edges and vertices does your house have?

- Does your house have any parallel faces? If so, how many?

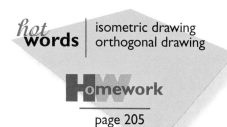

hot words | isometric drawing
orthogonal drawing

Homework
page 205

 **omework**

Planning and Building a Modular House

Applying Skills

How many faces does each structure have?

1.

2.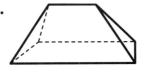

3.

4.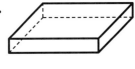

How many cubes are in each model? Remember that each cube on an upper level must have a cube below to support it.

5.

6.

7.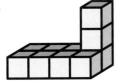

8.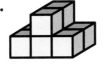

9. Of models 5–8, which ones have the same bottom layer?

Extending Concepts

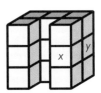

Look at this model house. Suppose you painted all the outside surfaces of the model, including the underside.

10. How many squares would you paint?

11. How many faces would be painted on the cube marked *x*? on the cube marked *y*?

12. Would any cube have 4 faces painted? 5 faces painted? Why or why not?

13. Draw a 6-face figure that is not a cube.

Writing

14. Look in the real estate section of a newspaper or in magazines about housing to find pictures that show houses or buildings in different ways. Find two or three examples of different ways to show structures. Tell why you think the artist chose each one.

Seeing Around the Corners

Applying Skills

Look at the first structure in each box. Tell which of the other structures are rotations of the sample. Answer *yes* or *no* for each structure.

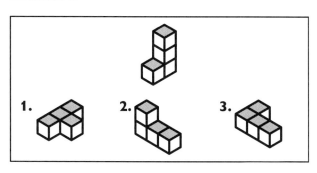

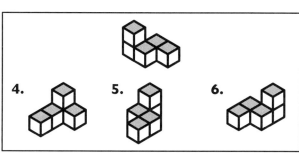

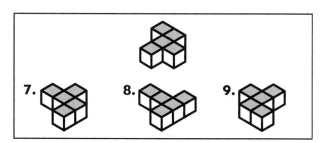

Extending Concepts

10. Draw this structure in a different position. Show how it would look:

  a. rotated halfway around

  b. lifted up, standing on its side

Making Connections

11. Draw the next structure in this pattern.

12. Make an isometric drawing of your house, school, or other building as if you were looking at it from the front. Then, make an isometric drawing of the structure as if you were looking at it from the left side.

Seeing All Possibilities

Applying Skills

Look carefully at each isometric drawing. The top side is shaded. Write *front, back, top, side,* or *not possible* for each orthogonal view.

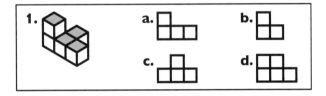

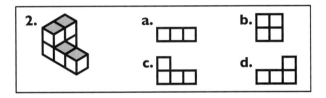

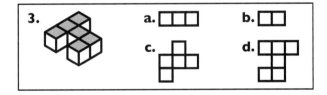

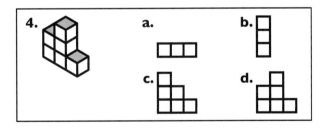

Extending Concepts

5. Make orthogonal drawings of this model that show these views:

   a. top view

   b. back view

   c. right side view

6. How would you tell a friend on the telephone how to build this structure?

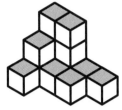

Writing

7. This dwelling is located in the Southwest of the United States. Why do you think it is shaped this way? What materials do you think are used? Give your reasons.

Picture This

Applying Skills

Look carefully at each set of orthogonal drawings. Choose which isometric drawings could show the same structure. Answer *possible* or *not possible*. Some structures may be rotations.

Set 1   Front   Side   Top
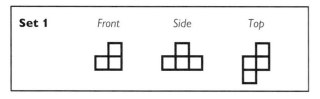

a.   b.   c.

Set 2   Front   Side   Top
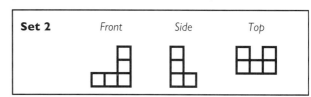

a.   b.   c.

Set 3   Front   Side   Top
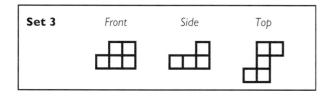

a.   b.   c.

Set 4   Front   Side   Top
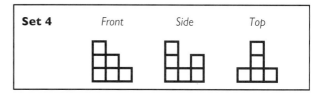

a.   b.   c.

Extending Concepts

Front view   Side view
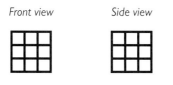

5. Draw at least three different top views of structures that all could have this front view and right side view.

Making Connections

6. Make orthogonal sketches of your school building. Show how you think it looks from the top, front, and right side.

String Shapes

Applying Skills

1. Which polygons have at least two parallel sides?

2. Which polygons have more than two pairs of parallel sides?

3. List all the equilateral polygons.

4. Which equilateral polygons have four sides?

5. List all the polygons with an odd number of sides.

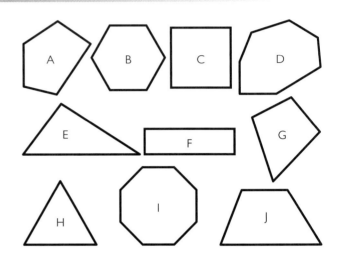

Extending Concepts

```
tri-
quad-
penta-
hexa-
hepta-
octo-
```

6. Choose a prefix for each of the polygons above that would help describe each figure.

7. Draw a different polygon to fit each prefix.

   a. Label the parallel and equal sides.

   b. Write clues to help someone identify each of your polygons.

Making Connections

8. Match the titles to the pictures. Tell why you made each choice.

   The Pentagon

   Tripod

   Hexagram

   Octopus

   Quartet

Polygon Paths

Applying Skills

For items 1 through 6, use a protractor and ruler.

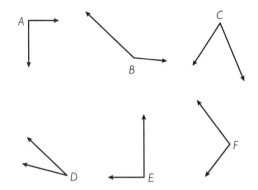

1. Which angles measure less than 90°?

2. Which angles measure greater than 90°?

3. Which angles measure exactly 90°?

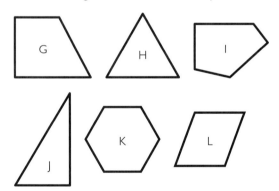

4. Which polygons have at least one right angle?

5. Which polygons have at least two equal angles?

6. Which polygons have all equal angles?

Extending Concepts

7. Use a protractor and ruler. Design and draw a path to create a polygon for each of the following specifications. What is the name of each polygon?

   a. four sides

      at least two sides parallel

      no 90° angles

   b. three sides

      one 90° angle

      two sides of equal length

Writing

8. Draw an imaginary neighborhood that includes a path made up of straight line segments (going from one end of the neighborhood to the other). Describe the path by giving the length of each line segment and the angle and direction of each turn that makes a path through the neighborhood.

   Example:

   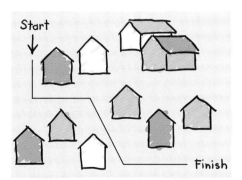

Shaping Up

Applying Skills

Tell whether each polygon is *regular* or *not regular*.

1.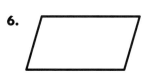

2.

3.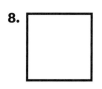

4.

5.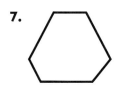

6.

7.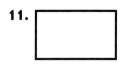

8.

Copy each quadrilateral and mark one pair of opposite angles.

9.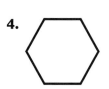

10.

11.

12.

Extending Concepts

String or rope can be used as a tool for making right angles.

Method: A twelve-foot length of rope is marked at every foot and staked into a triangle. The sides are adjusted until one side is 3 feet, one side is 4 feet, and the third side is 5 feet. When these sides are exact, the triangle will contain a right angle.

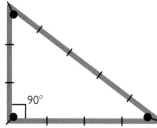

13. Find another trio of side lengths that will create a right triangle. Explain how you know you have a right angle. You can use string or rope to make a model. (An easy scale factor to use is 1 foot of rope in the problem equals 1 inch of string in your model.)

Making Connections

14. Find two other trios of side lengths that will create a right angle in a triangle. There are patterns to some of these trios of side lengths. If you see a pattern, describe it and make a prediction for how you would find other side lengths.

Assembling the Pieces

Applying Skills

1. Which of rectangles A–H have the same perimeter?

2. Which of rectangles A–H have the same area?

Use centimeter grid or dot paper.

3. Draw a polygon that is not regular and has the same perimeter as rectangle B.

4. Draw a regular polygon that has the same perimeter as rectangle F.

[Diagram showing rectangles A, B, C, D, E, F, G, H on dot paper]

Extending Concepts

5. Use a piece of string about 20 cm long. Knot it into a loop. Use the loop as the perimeter of at least three different triangles. Sketch your triangles on grid paper or dot paper to compare them. Estimate the area of each triangle.

6. Look over these definitions. Do they exactly describe each polygon? Can you improve the lists of clues?

   a. It has equal sides.
       It has parallel sides.
       It has four sides.

   b. It is not regular.
       Its sides are not equal.
       It has no parallel sides.

Writing

7. Answer the letter to Dr. Math.

> Dear Dr. Math:
>
> I don't understand how two rectangles with exactly the same perimeter can enclose different areas. Can you explain that to me?
>
> Perry Mitter

Beyond Boxes

Applying Skills

What two-dimensional shape will you see on the inside of each slice?

1.   **2.**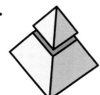

3.  **4.**

Copy the chart and fill it in for each structure.

5. **6.** **7.**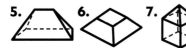

Shape of base			
Shapes used			
Number of faces			
Number of vertices			
Number of edges			
Sets of parallel faces			
Sets of parallel edges			

Extending Concepts

8. Draw three more three-dimensional figures. Add them to the chart you made for items **5–7.** Can you find a pattern in the relationship between the numbers of edges, vertices, and faces in the shapes? Tell all the steps in your thinking.

9. Can you draw a figure that does not follow the pattern you described in item **8**?

Writing

10. Imagine a penny. Now imagine ten pennies stacked carefully so that all the edges line up evenly.

a. Describe the shape of the stack. What properties does it have?

b. Imagine slicing that stack of pennies down the middle and opening the two halves. What is the two-dimensional shape of the new faces that are formed by the slice?

c. Think of another example of stacking a group of flat objects to form a new shape. Describe the new shape and how you would make it.

Drawing Tricks

Applying Skills

Is the figure a prism or a pyramid? Name the polygon forming the base.

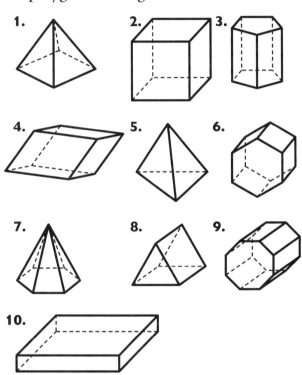

1.

2.

3.

4.

5.

6.

7.

8.

9.

10.

Draw each polyhedron. Tell how many faces it has.

11. Pentagonal pyramid

12. Pentagonal prism

13. Triangular prism

14. Triangular pyramid

Extending Concepts

15. Imagine that you are walking around the group of buildings in the pictures. Choose one picture as a starting point. List the pictures in the order that shows what you would see as you circle the buildings.

a.

b.

c.

d.

e.

f.

Writing

16. Tell how you visualized moving around the group of buildings.

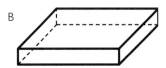

Mystery Structures

Applying Skills

For items 1–8, give the letters of the polyhedrons that fit the clues.

A 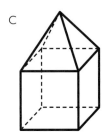   B

C   D

E   F

1. It has 7 vertices.

2. Its base is a square.

3. The structure is a prism.

4. It has exactly 3 sets of parallel faces.

5. All sides are equilateral.

6. It has 16 edges.

7. It has 4 sets of parallel faces.

8. It has no sets of parallel faces.

Extending Concepts

9. Draw two different polyhedrons with quadrilateral bases. Make a list of clues for each structure to help someone tell them apart.

Writing

My Mystery Structure
It has some edges.
All the edges are the same length.
It comes to a point on top.
Some of the pieces are triangles.
The base is a square.

10. Do you think this student wrote a good list of clues?

   a. Draw a structure that fits the clues.

   b. Could you draw a different structure that fits the same clues?

   c. Rewrite this list of clues if you think it can be improved. Explain your changes.

Putting the Pieces Together

Applying Skills

Make a chart like the one shown. Fill in the chart to describe each shape or structure.

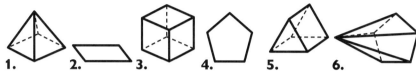

1.   2.   3.   4.   5.   6.

Name of shape or structure						
Polygon or polyhedron						
Number of sides or faces						
Number of vertices						
Number of right angles						
Sets of parallel sides or edges						

Extending Concepts

Using the pieces shown, tell how many of each piece you would need to make each polyhedron. (Assume that you can make matching edges fit.)

7. Hexagonal pyramid   **8.** Triangular prism

9. Cube   **10.** Pentagonal prism

Writing

11. Write a letter giving some good advice for next year's designers. Tell about some things that helped you with the investigations.

Glencoe

The McGraw-Hill Companies

This unit of MathScape: Seeing and Thinking Mathematically was developed by the Seeing and Thinking Mathematically project (STM), based at Education Development Center, Inc. (EDC), a non-profit educational research and development organization in Newton, MA. The STM project was supported, in part, by the National Science Foundation Grant No. 9054677. Opinions expressed are those of the authors and not necessarily those of the Foundation.

CREDITS: Unless otherwise indicated below, all photography by Chris Conroy.

162 Images; **163** (tl tr)Images; **164** Photodisc/Getty Images; **164** (tc)Slyvian Grandadam/Photo Researchers; **165** (t)Norris Taylor/Photo Researchers, (c)Andrea Moore; **167** Norris Taylor/Photo Researchers; **174** Photodisc/Getty Images; **179 through 193** bar art excerpted from Shelter © 1973 by Shelter Publications, Inc., P.O. Box 279, Bolinas, CA 94924/distributed in bookstores by Random House/reprinted by permission. **184** (background)Images, (foreground)Photodisc/Getty Images.

Send all inquiries to:
Glencoe/McGraw-Hill
8787 Orion Place
Columbus, OH 43240-4027

ISBN: 0-07-866798-4

5 6 7 8 9 10 058 13 12 11 10 09 08